Science Crossword Puzzles
Grades 3–6

```
H U M A N B O D Y
      N
      I
      M A R I N E L I F E
      A
      L           M       M
      C L A S S I F I C A T I O N
E A R T H         T       T
      A           T       I
      E N E R G Y         E       O
      A           R       N
      F O R C E
      T           R   P
      W E A T H E R   O   L
      R           C   A
      I           K   N
      S O L A R S Y S T E M
      T               T
      M I N E R A L S
      C
P L A N T S
```

Written by Rebecca Stark

ISBN 978-1-56644-566-5

Educational Books 'n' Bingo

Formerly published by Educational Impressions, Inc.

Printed in the United States of America.

TABLE OF CONTENTS

*An alphabetical list of possible answers from which to choose is provided for each crossword puzzle. Use these lists at your discretion.

Animal Characteristics

ACROSS

5. Warm-blooded, feathered vertebrate
8. Cold-blooded vertebrates with scales
9. Dogs, wolves, coyotes and foxes
11. Internal framework of vertebrates, which is made of bone or cartilage
12. Animals that maintain a constant body temperature (2 words)
13. Basic unit of biological classification
14. Animals with an endoskeleton; amphibians, birds, fish, reptiles, mammals
17. Vertebrates that undergo metamorphosis, such as frogs and salamanders
19. Insects and other invertebrates with a segmented body and jointed appendages
21. Animal without a backbone
22. They get all their energy from eating plants
27. Lions, tigers, and domestic cats are examples
28. A carnivorous mammal with flippers; a walrus is one
30. Snails, mussels, clams and octopuses are examples

DOWN

1. Females carry their young in a pouch
2. Complete this analogy: bird : biped :: mammal : ___
3. Animal that eats meat
4. Organs of respiration; fish and young amphibians have them
6. Arthropod whose adult stage has three pairs of legs
7. Spiders and scorpions, for example
10. Complete ___ has 4 stages
15. Aquatic vertebrate with scales and gills
16. Mammals with advanced binocular vision, such as humans, monkeys and apes
18. An animal that hunts and eats other animals
20. Sea mammals including whales, dolphins and porpoises
23. Horses, zebras and donkeys
24. Cows, oxen and buffalo
25. Get their food from both plants and animals
26. Beavers and other vertebrates with continuously growing incisor teeth
29. Those of birds are enclosed in a chalky shell; those of reptiles are in a leathery membrane
30. Warm-blooded vertebrate with hair

Animal Characteristics

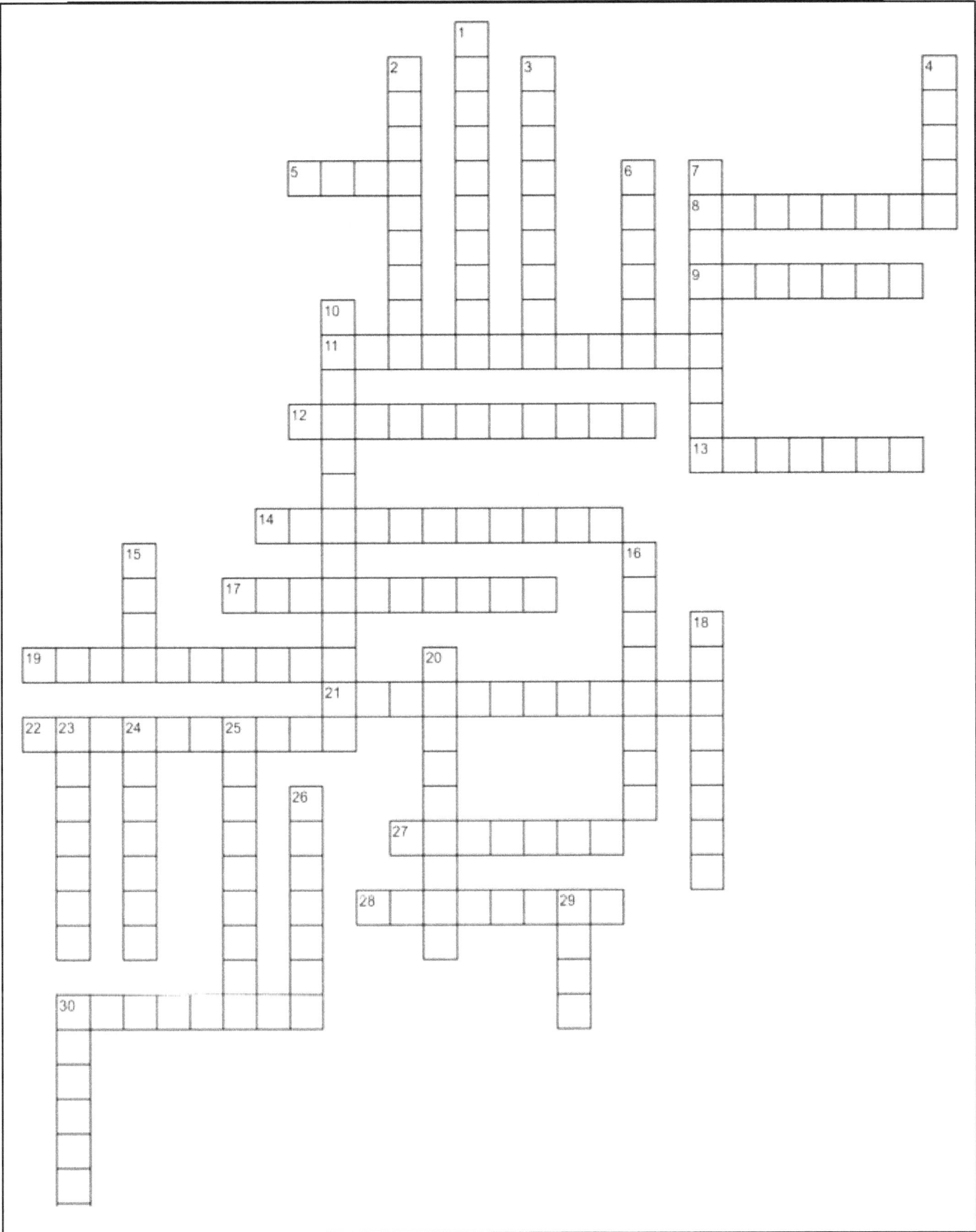

Matter and Energy

ACROSS

4   Transfer of thermal energy from one substance to another

6   Amount of matter in an object

8   Elements are made up of them

9   ___ energy can cause movement of a charged particle through a wire

11 Material that is a poor conductor

13 Change of a liquid into a gas

14 Anything that has mass and occupies space

16 Ability to do work or produce a change

17 Type of energy acquired by the objects upon which work is done

19 Container in which chemical energy is converted into electricity

21 Metal is a good ___ of electricity

24 State of matter which has definite volume but no definite shape

25 Only form of energy that we can actually see directly

26 The force of gravity on the object; how heavy something is

27 Produced when a force causes an object or substance to vibrate

DOWN

1   As wood burns, ___ energy is converted into heat and light

2   A closed path through which an electric current flows

3   Energy a body possesses because of its motion

5   An electric ___ is the measure of the extra positive or negative particles

7   Wind, water and solar energy are this type of energy

10 Antonym of "evaporation"

12 Stored energy in an object

15 State of matter that has no fixed shape or volume

18 Substance made from a single type of atom

20 Energy produced by the sun

22 Refers to how close together the molecules of a substance are

23 State of matter with a definite shape and a definite volume

26 The transfer of energy

Matter and Energy

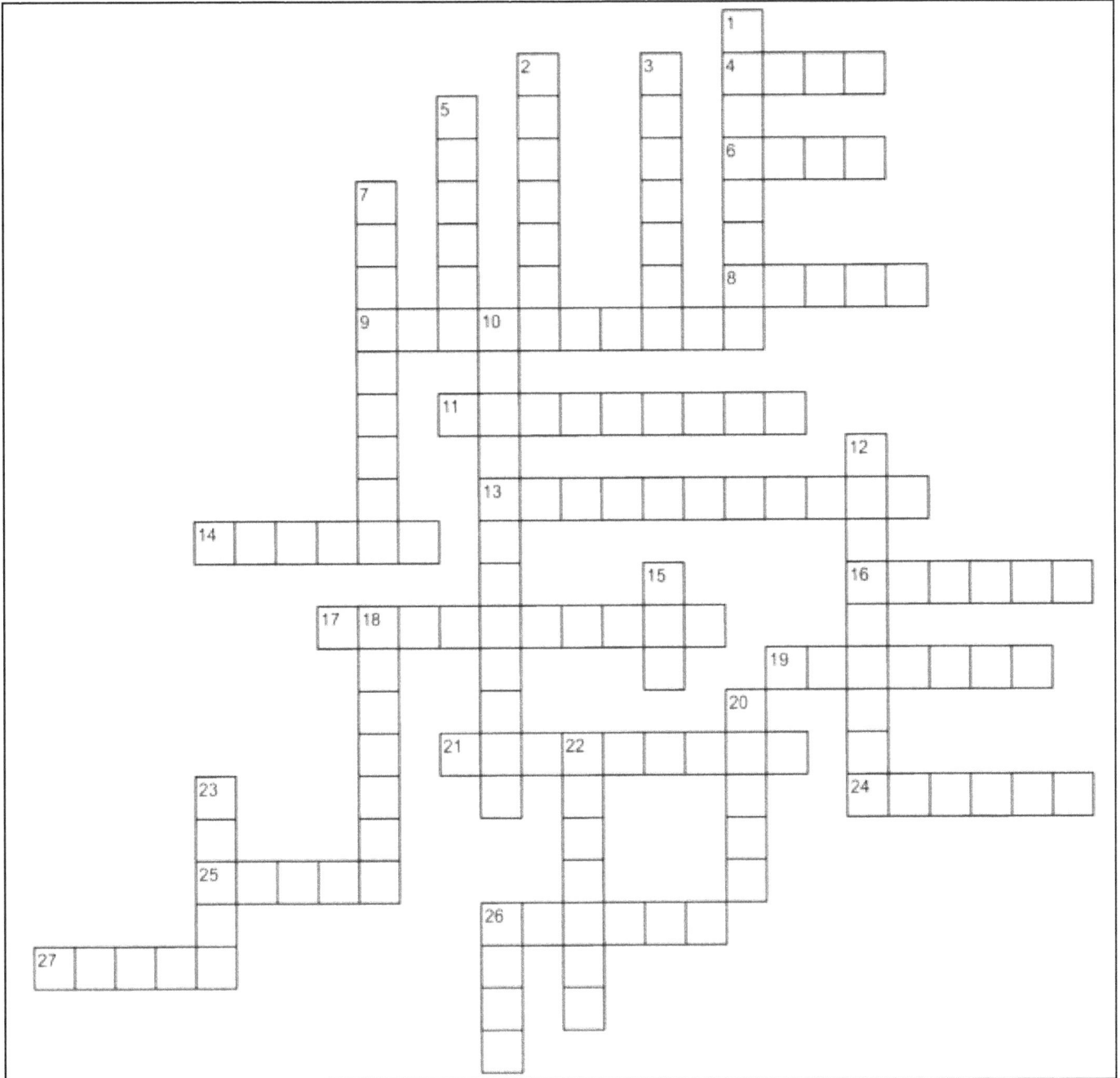

Forces and Motion

ACROSS

1. Anything that has mass and takes up space
5. Movement of an object from one place to another
6. Actually a lever that rotates in circle around a fulcrum (3 words)
10. Substance used to reduce the friction between two solid surfaces
12. Simple machine consisting of a rigid bar that can pivot on a fulcrum
14. Force that acts in the opposite direction of a moving object
15. Force tending to stretch or elongate an object
17. Strength or force that something has when it is moving
18. Rate of change in velocity, or speed
19. Simple machine thick at one edge and tapered to a thin edge at the other
21. A toothed wheel
22. Describes stored energy an object has because of its position or state
25. Amount of matter in an object
27. A ramp is this kind of simple machine (2 words)
29. Describes a machine that combines 2 or more simple ones
30. Measure of how strongly gravity pulls on matter, or heaviness

DOWN

1. Mechanical device that transmits, modifies, or changes the direction of force
2. Inclined plane wrapped around a shaft
3. A push or pull that causes an object to move, stop or change direction
4. Describes energy in motion or use
7. Magnetism that is produced by an electric current energy
8. Force that resists motion when two surfaces rub together
9. What an object less dense than the fluid it is in will do
11. Measure of how quickly work is done
13. Heat, light, sound, electrical, and wind are some forms
16. How fast something is moving
17. Object that attracts iron
20. Weight to be carried or moved
23. Force of attraction between objects
24. Resistance of a physical object to a change in its state of motion
26. Force used by vacuum cleaner to clean carpets
28. Simple machine that uses grooved wheels and a rope
30. Force used to move something

 Science Crossword Puzzles: Grades 3 & Up

Forces and Motion

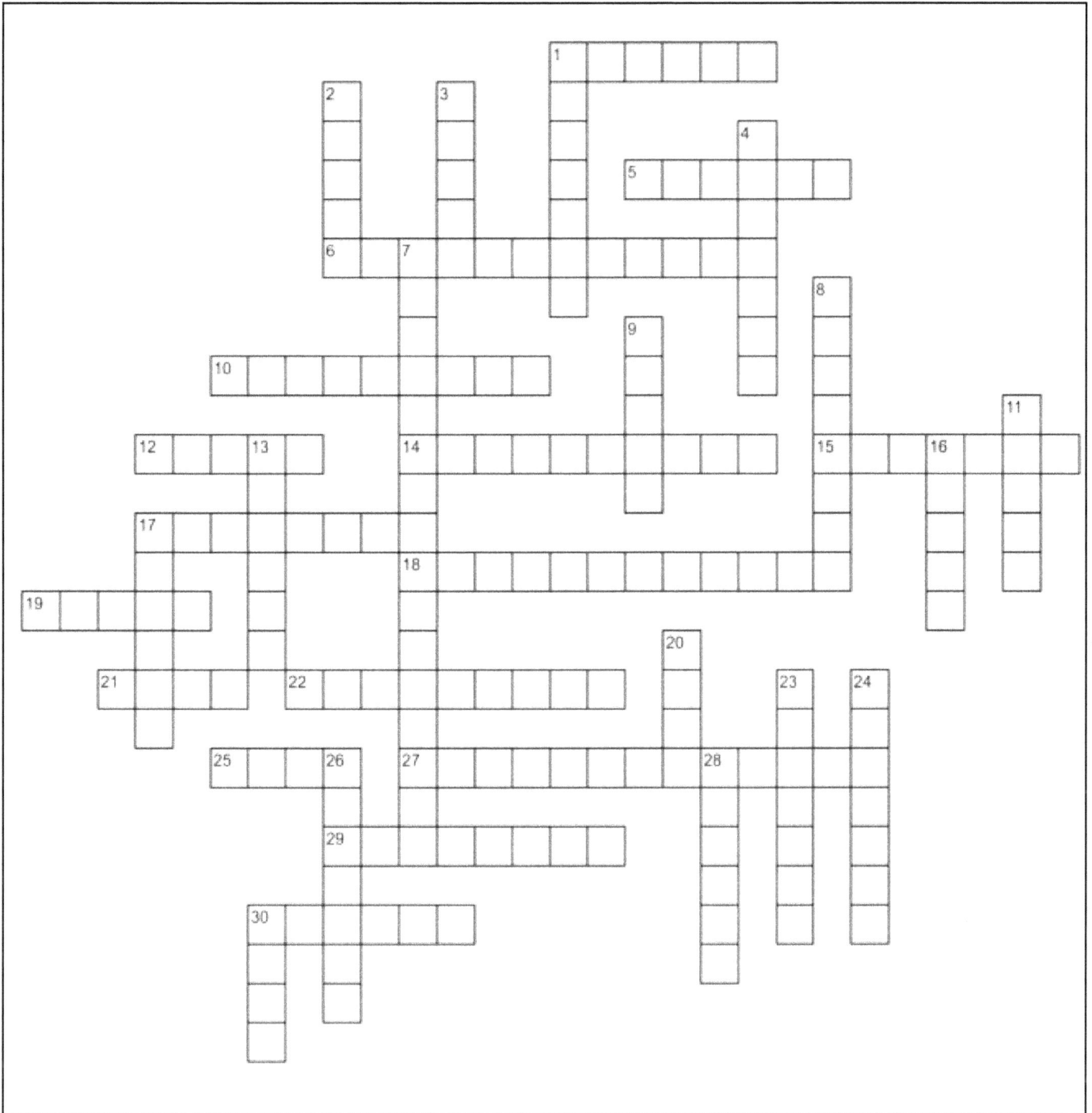

The Human Body

ACROSS

2. Large blood vessel
4. System that rids body of toxic waste; part of excretory system
6. System that includes bones and the tissues that connect them
9. Fluid circulated through body by heart
10. Bronchi and lungs and part of the central ___ system
12. Blood vessels that carry blood toward heart; not as strong as arteries
13. Smooth, cardiac and skeletal are 3 types
14. Adult human body has 206
15. Basic structure and functional unit of all living things
16. Thin, tubular bundle of nervous tissue along back
17. ___ arteries carry blood from heart to lungs
18. Filter wastes from the blood; most people have 2
24. Group of cells that perform a specific function
25. Part of skeletal system; protects brain and eyes; called cranium
26. Organ that controls all we do
27. Made up of organs and related tissues concerned with the same function
29. Muscle that sends blood around the body

DOWN

1. Brain and spinal cord make up the central ___ system
3. Mouth, esophagus, stomach, small intestine, and colon are part of this system
5. Capillaries are smallest vessels in this system
7. Sometimes called the voice box.
8. Shoulder, hip, knee, and ankle
11. This organ of smell is part of respiratory system
19. System that helps keep out harmful bacteria and viruses
20. Detects, processes and sends sound signals to brain
21. Sac-like organ; main organ of digestion
22. Basic unit of heredity
23. Basic respiratory organs; remove carbon dioxide from blood and bring oxygen into it
28. Their outer layer is made of enamel, the hardest substance in the body
30. Organ of sight

The Human Body

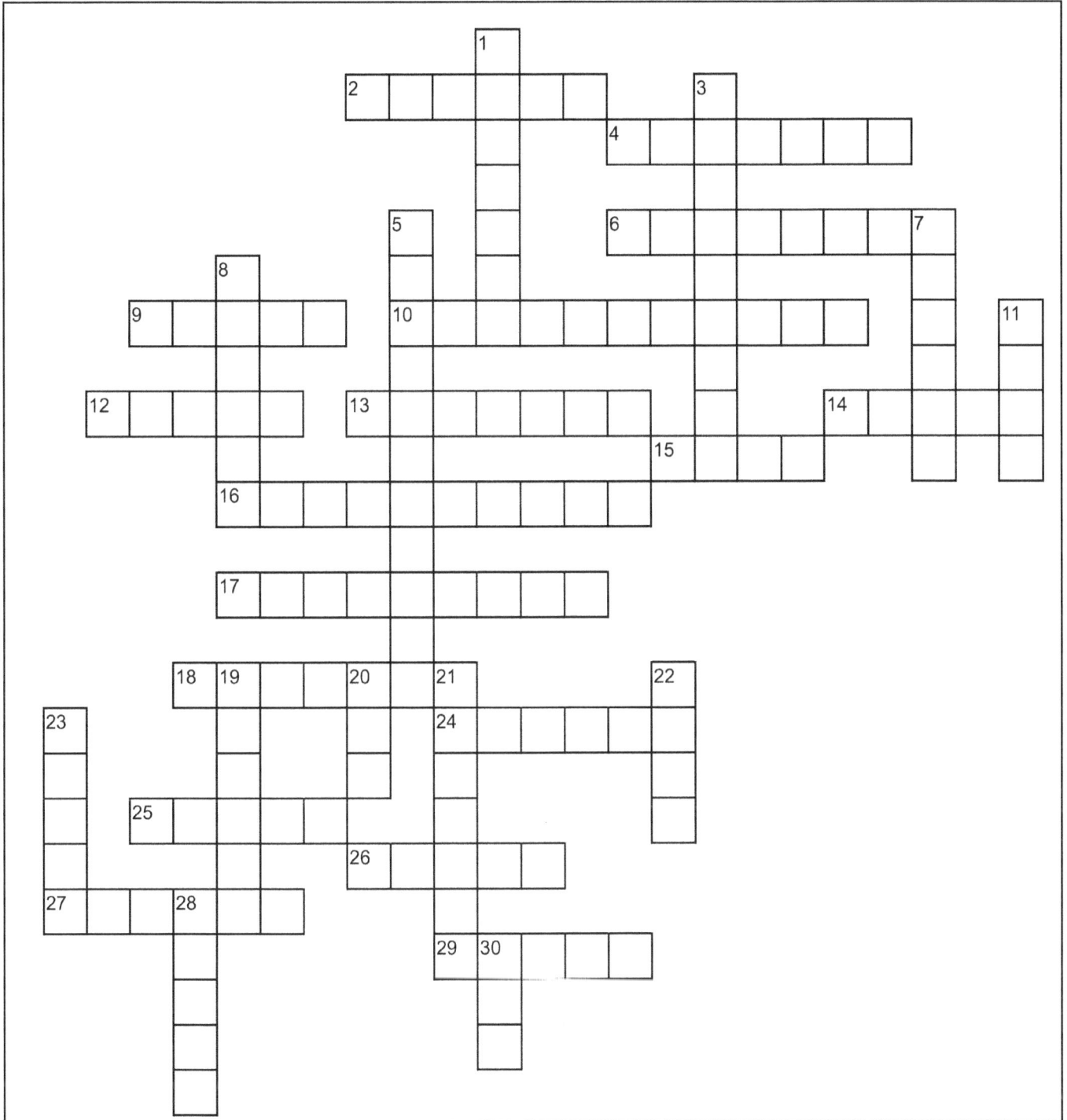

Marine Life

ACROSS

3. Seals, sea lions and walruses
7. Porous primitive life forms; water flows in and out of their bodies
10. Respiratory organs of sea mammals
12. Large aquatic, flightless bird of Antarctica (2 words)
13. Largest animal in the sea (2 words)
14. Can release a dark ink, clouding the water
15. Lobsters and crabs, for example
16. Free-swimming aquatic animals with a soft body
18. A fish with a skeleton of cartilage
19. Scientific study of the ocean
23. Dolphins, seals, and manatees
28. Underwater ecosystem built by colonies of tiny animals (2 words)
29. Vertebrate with gills
30. Alternating rising and falling of the surface of the ocean

DOWN

1. Large body of salt water
2. Respiratory organs of fishes
4. Biological community of interacting organisms
5. Microscopic plants and animals that drift in the water
6   Water here is sometimes salty and sometimes freshwater
8. Unlike other species, the___ cannot retract its legs and head into its shell (2 words)
9. Animal that eats other animals; killer whale is one
11. Largest living thing on Earth; visible from outer space (3 words)
17. Same kind of fish traveling together
20. Adjective meaning "of, found in, or produced by the sea"
21. Has five or more radiating arms; also called sea star
22. A small free-swimming crustacean with an elongated body
24. Sometimes called a sea cow; breathes air at surface of water
25. Has a soft, unsegmented body enclosed in a shell; a clam is one
26. Refers to the concentration of salt in the ocean
27. This mammal is a cetacean, a carnivore with fins

Marine Life

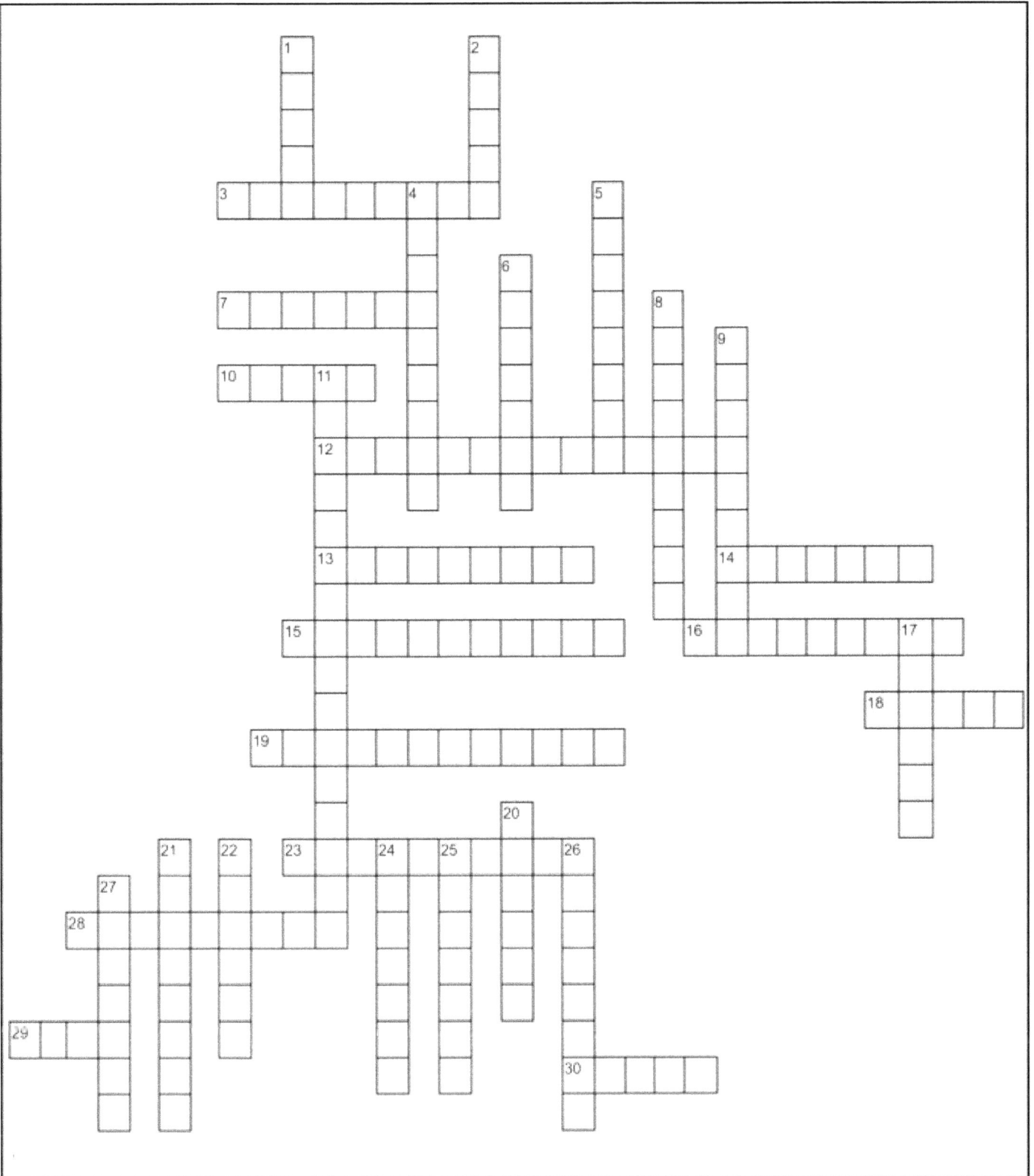

Our Solar System

ACROSS

5. Complete this analogy: sun : solar :: moon : ___
6. Crew member in a spacecraft
8. Huge ball of gases held together by gravity that radiates energy
9. Mass of gas surrounding a planet
10. Branch of science that deals with celestial objects
12. At the center of our solar system; a medium-sized yellow star
13. Relating to the sky or outer space
14. Comprises all the space and matter in existence
20. Small bodies that travel through space
22. A planet's natural satellite
24. Third planet from sun
25. Path of a heavenly body around another body planet
26. Haley's ___, a famous one, is seen from Earth every 76 years

DOWN

1. Fifth from sun and largest planet; has great red spot
2. Ice giant named after ancient Roman god of the sea; 8th planet from sun
3. Like Neptune, this 7th planet from the sun is an ice giant
4. Acronym for National Aeronautics and Space Administration
5. Distance light travels in a year (2 words)
7. The sun, the planets, and the other celestial bodies that orbit the sun
11. Named after ancient Roman god of war; fourth planet from sun
12. Planet known for its spectacular planetary rings
15. Second planet from sun; named after ancient Roman goddess of love
16. Closest planet to the sun
17. Planets orbit the sun because of this force
18. Our solar system is in this galaxy (2 words)
19. One of the small metallic and rocky bodies that orbit the sun
21. A solar one occurs when the moon passes between the sun and Earth
23. In 2006 its status was downgraded to "dwarf planet"

Our Solar System

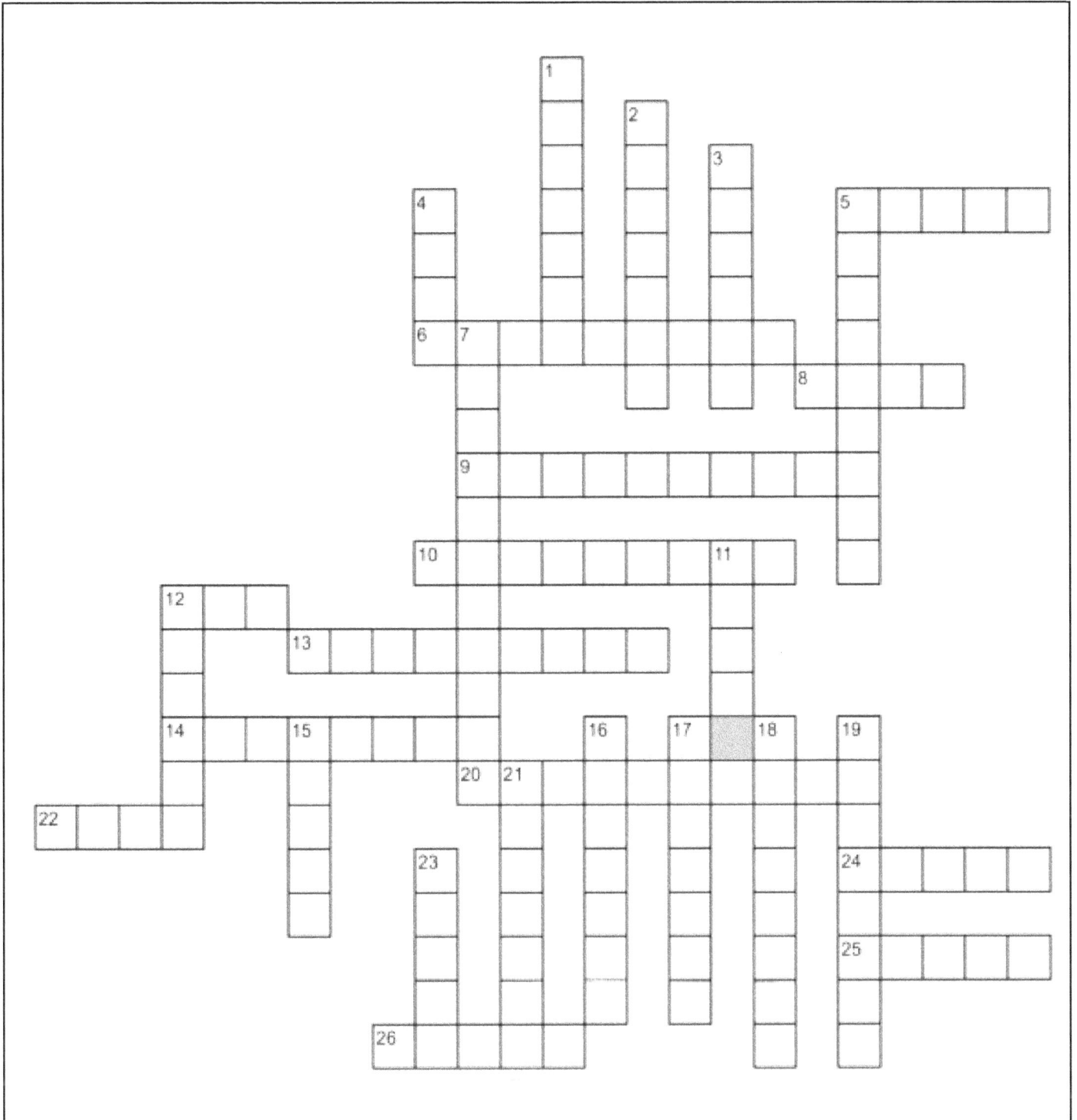

Plants

ACROSS

3. Flowerless plant with roots; stems; and large leaves, called fronds
5. They anchor the plant and absorb water and minerals from the soil
7. Reproductive structure of a plant
10. Cereal grass
11. Transfer of pollen from anther to stigma of a flower
13. Sweet liquid produced by some flowers
14. Plants with weak stems; need support as they grow
15. Gas used by plants during photosynthesis (2 words)
16. Trees with cones and evergreen leaves
17. Plant part that manufactures food through photosynthesis
21. Plant that lives for more than two years
22. Provides support for the plant and holds it up to the light
25. Pigment that absorbs light and allows photosynthesis to occur
26. Leaves, stems, roots, or other parts of certain plants that people eat
30. Watering of land by artificial means
31. Trees that shed foliage at the end of the growing season
33. Resting stage of a plant that is formed underground
34. Small, flowerless green plant; grows in tufts on moist ground, tree trunks, etc.

DOWN

1. Female, ovule-bearing part of the flower
2. Male reproductive organ of flower; bears fine powder called pollen
3. Ecosystem dominated by trees
4. Kingdom to which plants belong
6. Plants valued for their flavor, scent or medicinal value
8. Plant that completes its life cycle in a single year
9. Thin upper layer of earth in which roots of plant grow
11. Process by which plants use water, light and carbon dioxide to make their own food
12. Shoot or bud that has been joined to another plant
18. In the fruit of a plant; they are dispersed in many ways
19. One of the modified leaves that surround the reproductive part of a flowers
20. Low shrub with a lot of branches
23. A woody perennial plant with a single main stem, called a trunk
24. Hard-shelled seed
27. Edible reproductive body of a seed plant
28. Succulent, spiny plant
29. Branch of biology that deals with plant life
32. Basic unit of life
33. Small bulge on a stem or branch with an undeveloped shoot, leaf, or flower

Plants

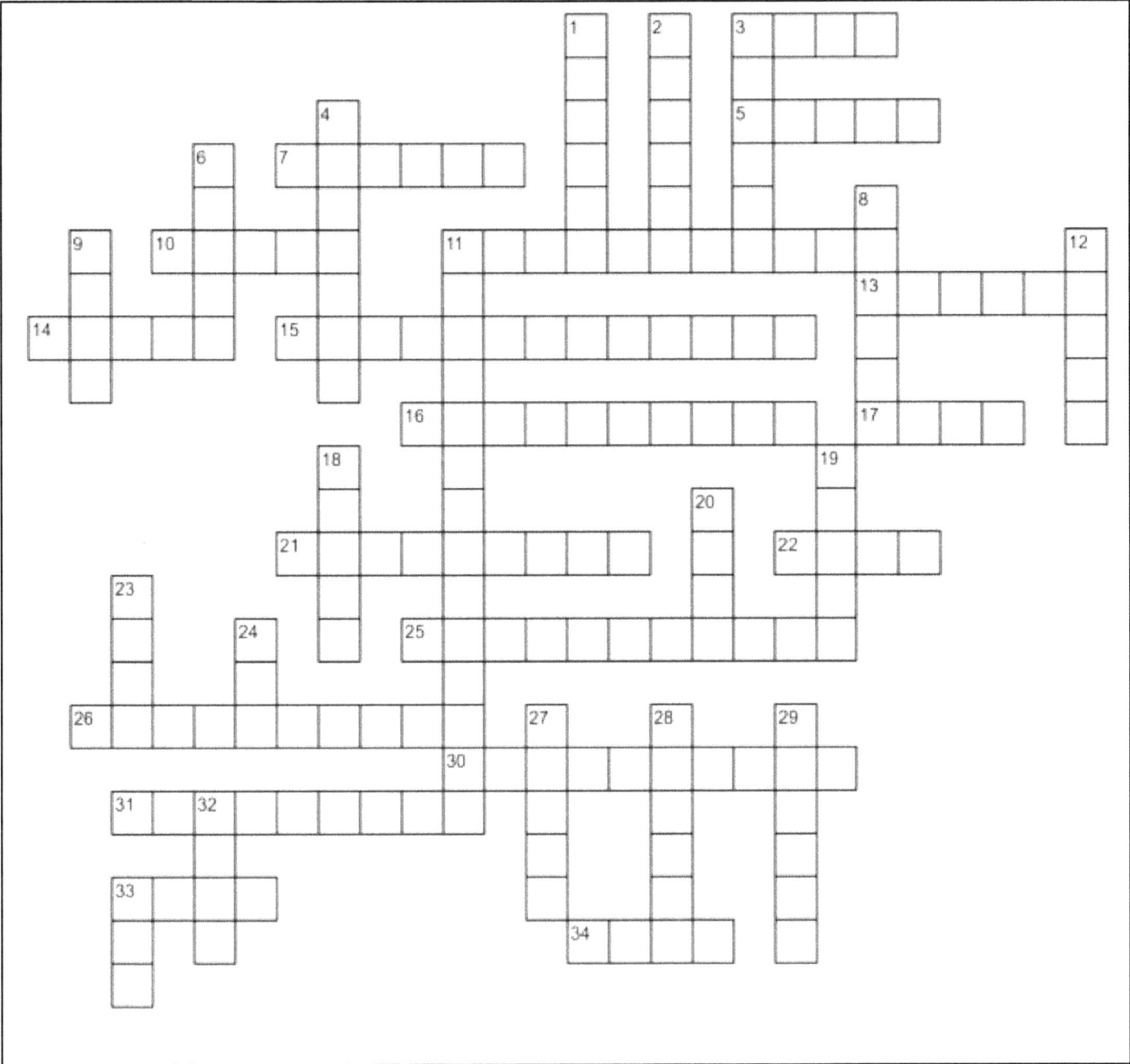

Weather

ACROSS

6. Storm characterized by funnel-shaped cloud
8. Formed by layers of clear ice and compact snow
9. Clouds that appear as a horizontal layer of grey
10. Degree of hotness or coldness
14. Evaporation, condensation, precipitation and collection (2 words)
18. Process of changing from a gas into a liquid
19. High, white, feathery clouds
20. Boundary between two different air masses
21. Winds are named by the ____ from which they come
22. Process of changing from a liquid into a gas
23. Severe storm with winds of 75 miles per hour and greater
25. Electric discharge in the atmosphere
27. Moving air
28. A wind with speeds of 4 to 7 miles per hour is a light one

DOWN

1. Average weather over a period of time
2. A severe winter storm
3. A heavy rain
4. ____ : rain :: flurries : snow
5. Droplets of water vapor suspended near the ground
7. Extended period of time without rain
9. Precipitation in the form of 6-sided ice crystals
11. Rain, snow, sleet and hail
12. Dampness in the air
13. What the prefix "nimbo" or suffix "nimbus" designates (2 words)
15. Temperatures decrease the further away from it you go
16. Puffy clouds
17. Water droplets that condense on cool surfaces
24. Condensed vapor in the atmosphere
26. Loud sound produced by expanding air

Weather

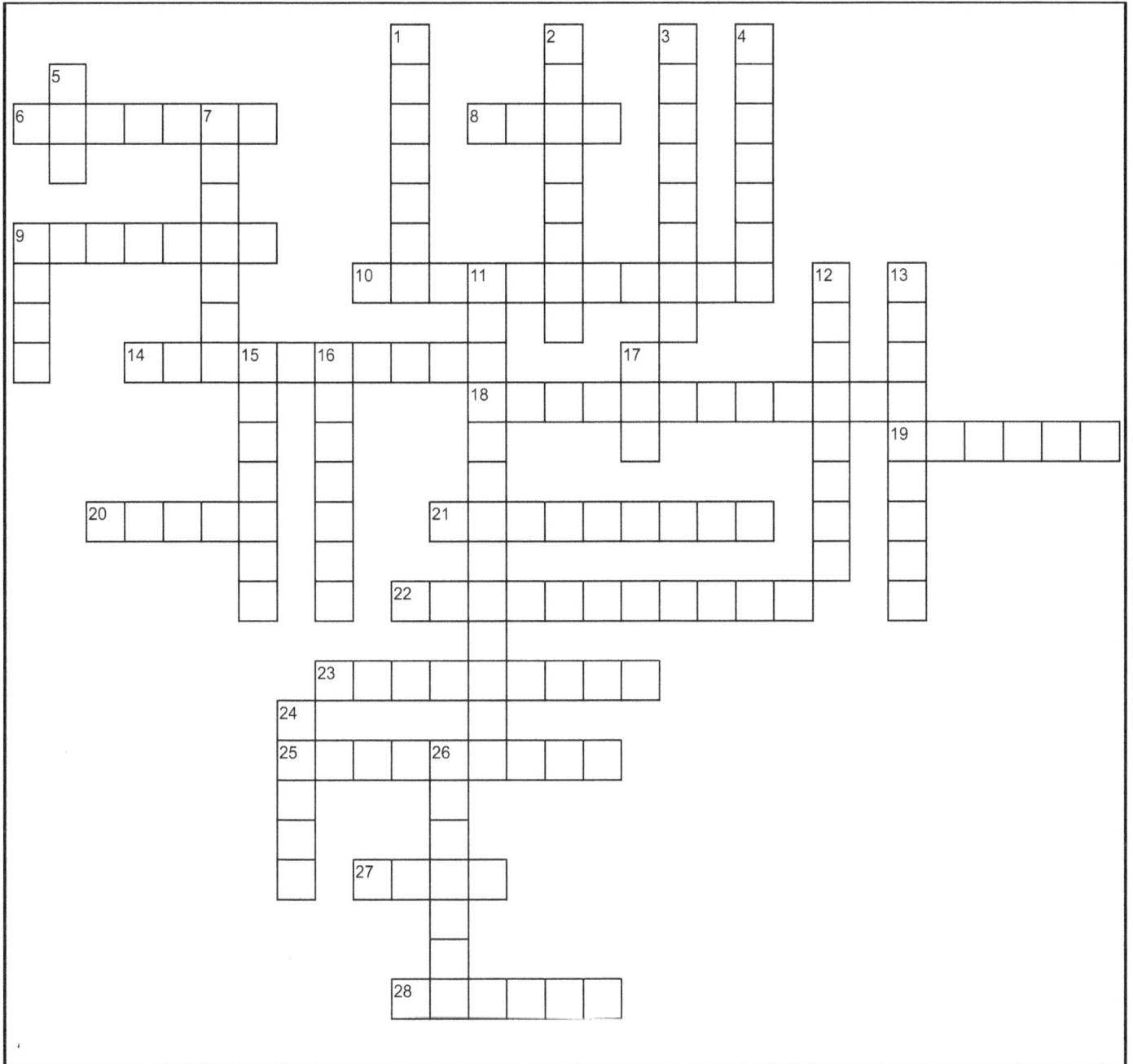

Earth Science

ACROSS

2. Long, narrow, steep-sided depressions in the ocean floor
5. A natural pool of hot water that sometimes erupts and gushes into air
8. Vent in Earth's crust through which lava, steam, ashes, etc., are expelled
9. Salt content of a body of water
11. Long, high sea wave caused by an underwater earthquake or other disturbance
12. Geological formation containing or conducting ground water
13. Science that deals with Earth's physical structure and substance
16. Scientist who studies weather and climate
17. Scientist who studies volcanoes
18. Used to rate the strength or total energy of earthquakes (2 words)
20. Height above sea level
21. Outermost layer of a planet
23. Part of atmosphere where weather takes place
25. Gases surrounding the earth or another planet
28. Vast body of salt water that covers almost 3/4 of Earth's surface
29. Describes an object that's reduced to liquid form by heating
30. Between the crust and the core.
31. Relating to earthquakes

DOWN

1. Central part of the Earth and other rocky planets
3. Sudden movement of the Earth's lithosphere
4. Breaking of rock into smaller pieces by water, wind, and ice
6. One of the seven large landmasses on Earth
7. Crack in Earth's crust
10. Submerged ridge of rock or coral near the surface of the water
14. Molten rock that has reached the surface
15. Huge, movable segments of Earth's crust (2 words)
19. Fine particles of mineral matter from a volcanic vent
21. Continuous movements of a fluid body, such as water or air, in a certain direction
22. Average weather conditions for an area over a long period of time
24. Abrupt downhill movement of soil and bedrock
26. Movement of rocks and soil to other places due to weathering
27. Burst forth
32. Molten rock beneath Earth's surface

Earth Science

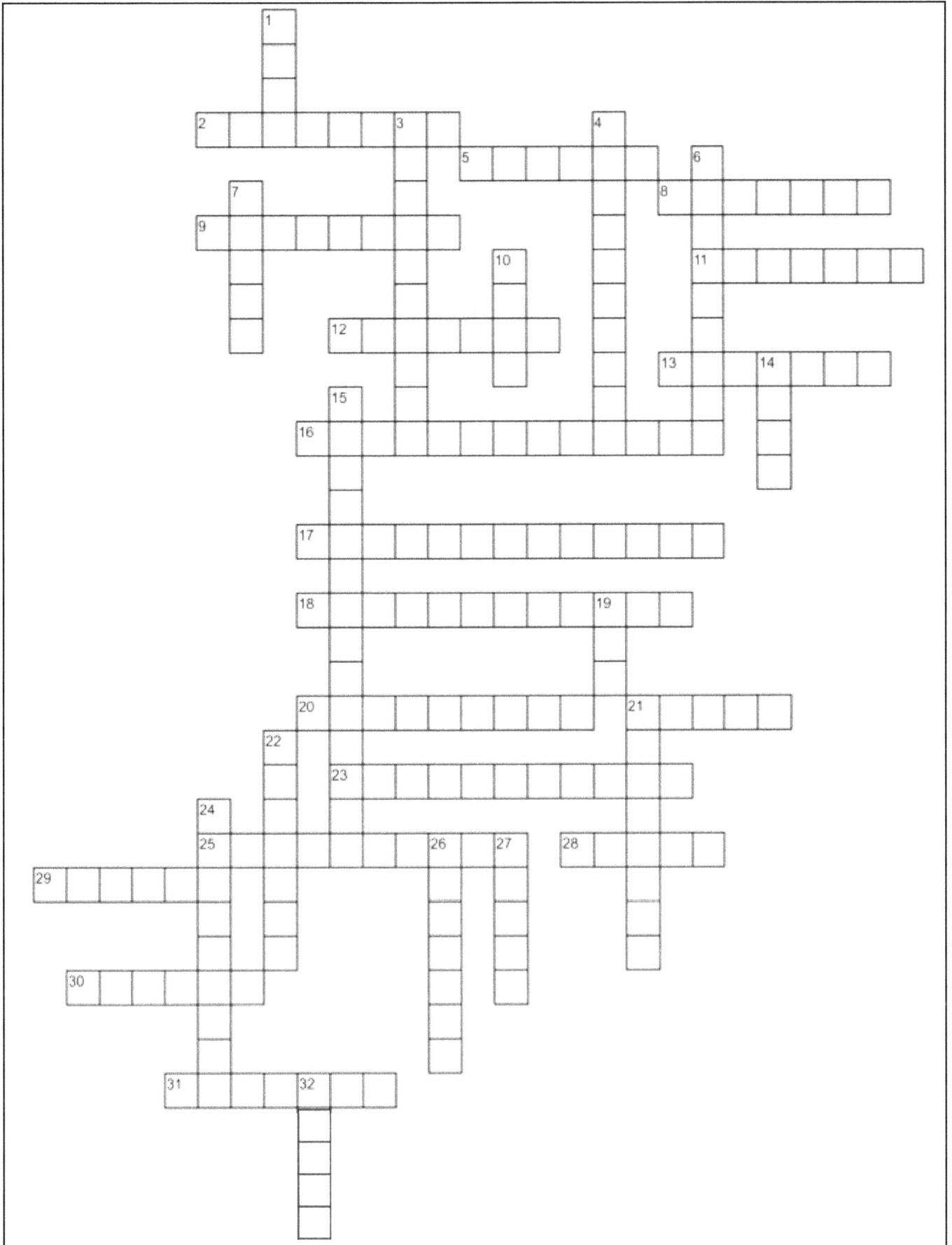

Rocks and Minerals

ACROSS

2. Remains of ancient animals and plants

4. Science that deals with the earth's physical structure and the processes that act on it

6. Type of rock that solidified from lava or magma

7. What layers of sedimentary rocks are called

10. Rock formed by sediment deposited over time by water or air

11. Formation that hangs from the ceiling of a cave

12. Ruby and sapphire, for example

18. Hardness, streak, fracture, color and luster are ___ of minerals

19. Formation that grows from the cave floor

20. Metamorphic rock formed from limestone

22. Yellow precious metal

23. Iron, gold, silver, copper, and aluminum

24. Naturally occurring solid substance that is not of plant or animal origin

26. Molten rock beneath Earth's surface

27. Color of mineral's powder when dragged across unglazed piece of porcelain

28. Sold rock beneath loose material such as soil and sand

DOWN

1. A hard sedimentary rock

3. Essential constituent of most crystalline rocks

5. Dark, fine-grained volcanic rock

8. Rock changed to another form due to heat, pressure, or another agent

9. A red-brown metal

13. What Mohs Scale measures

14. Adjective describing shiny chemical elements that conduct heat or electricity and can be formed into sheets

15 Also known as volcanic glass

16. Metamorphic rock formed from sandstone, a sedimentary rock

17. Hardest naturally occurring substance

21. Second most common mineral

25. Molten rock after reaching Earth's surface

29. Made up of two or more minerals

30. What a spelunker explores

Rocks and Minerals

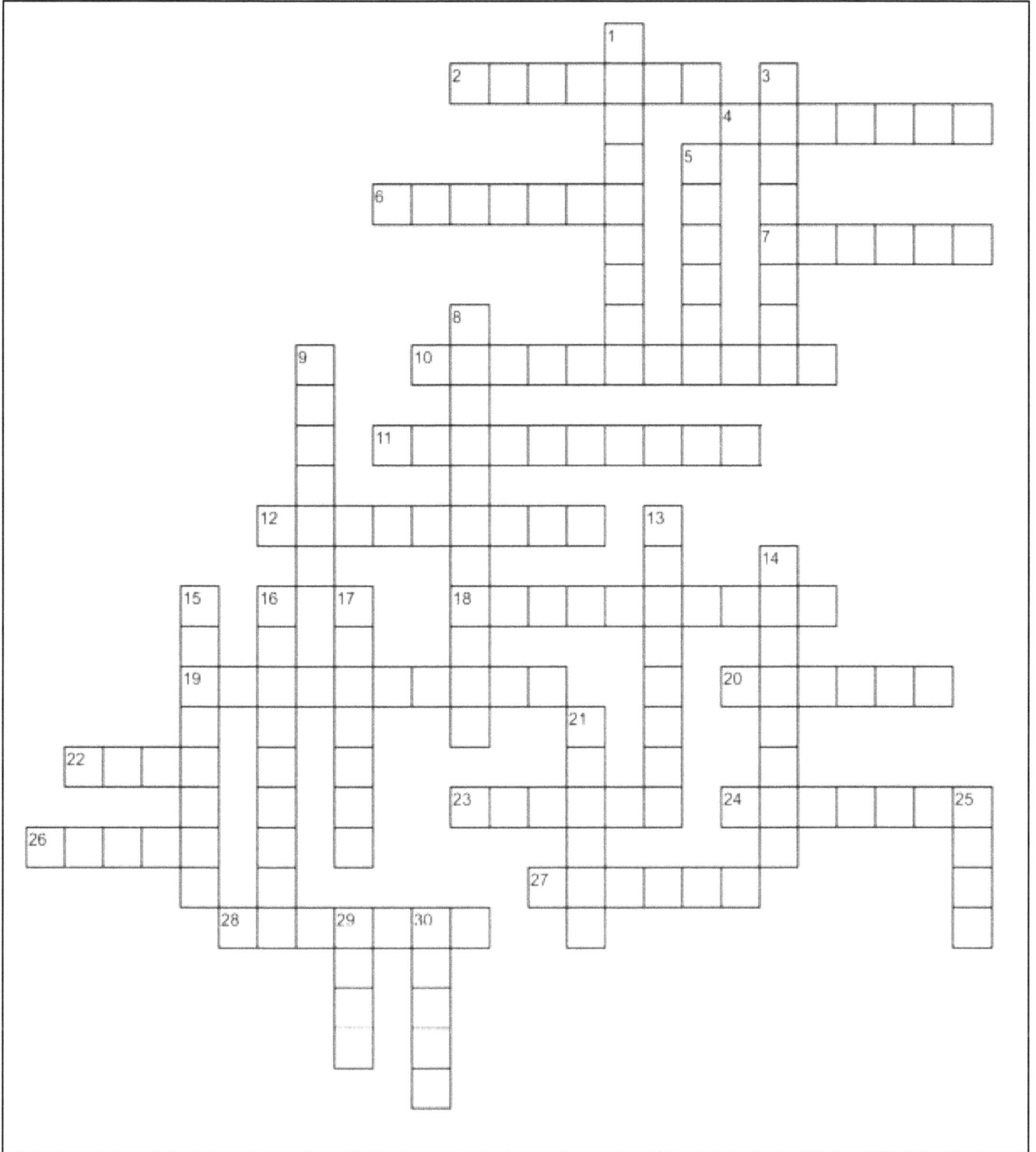

Classification of Living Things Word Search

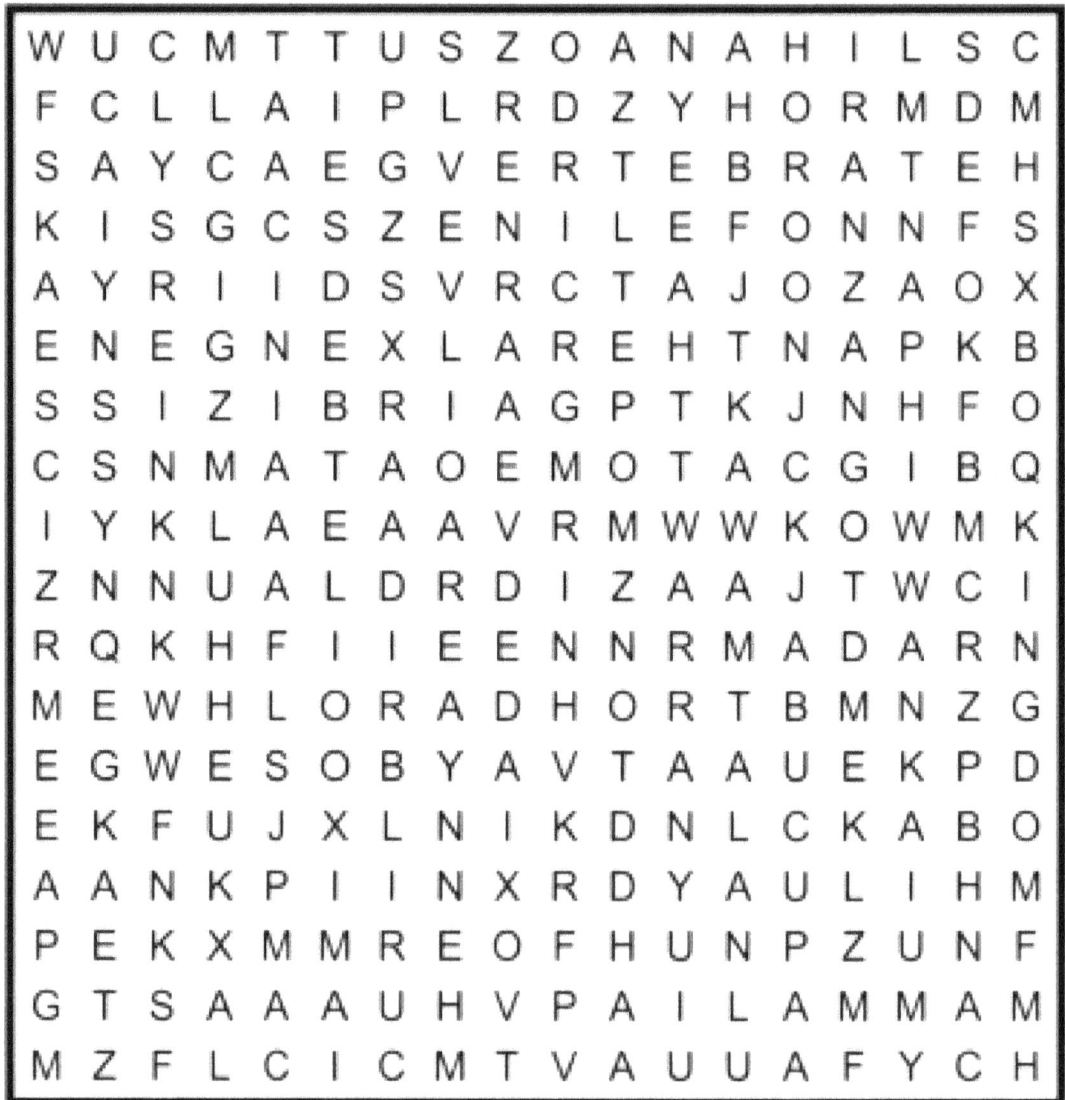

```
W U C M T T U S Z O A N A H I L S C
F C L L A I P L R D Z Y H O R M D M
S A Y C A E G V E R T E B R A T E H
K I S G C S Z E N I L E F O N N F S
A Y R I I D S V R C T A J O Z A O X
E N E G N E X L A R E H T N A P K B
S S I Z I B R I A G P T K J N H F O
C S N M A T A O E M O T A C G I B Q
I Y K L A E A A V R M W W K O W M K
Z N N U A L D R D I Z A A J T W C I
R Q K H F I I E E N N R M A D A R N
M E W H L O R A D H O R T B M N Z G
E G W E S O B Y A V T A A U E K P D
E K F U J X L N I K D N L C K A B O
A A N K P I I N X R D Y A U L I H M
P E K X M M R E O F H U N P Z U N F
G T S A A A U H V P A I L A M M A M
M Z F L C I C M T V A U U A F Y C H
```

All living organisms are classified into groups
based on basic, shared characteristics.
For the word search, find the words in each list.

Seven Levels of Classification	Classification of a Tiger: Scientific Names	Classification of a Tiger Common Names
Kingdom	Animalia	Animal
Phylum	Chordata	Vertebrate
Class	Mammalia	Mammal
Order	Carnivora	Carnivore
Family	Felidae	Cat (Feline)*
Genus	Panthera	Big Cat
Species	*Panthera tigris*	Tiger

*Include both in search.

Science Crossword Puzzles: Grades 3 & Up

Hidden Anagrams: Science Theme

Find the anagram of a science term hiding in each sentence.

1. A deep moat surrounded the castle.
CLUE: Smallest unit of an element that retains the chemical properties _____

2. The teen searched in the couch cushions for the remote control.
CLUE: Small body of matter from outer space that enters the earth's atmosphere _____

3. The shelf seemed to sag in the middle because of the heavy weight placed on it.
CLUE: Substance that is like air and has no fixed shape _____

4. The boy began to lament his decision not to attend the party.
CLUE: Beyond the Earth's crust _____

5. The speech given during the assembly was very topical.
CLUE: Relating to or using light _____

6. The guest asked her hostess for the recipe for the clam dip.
CLUE: Pleasantly free from wind _____

7. The young child was learning how to tie the laces on his shoes.
CLUE: Small, thin horny or bony plate protecting skin of fish and reptiles _____

8. The chef used many different oils, but olive oil was his favorite.
CLUE: The upper layer of earth in which plants grow _____

9. The directions on the treasure map said to take three paces to the right.
CLUE: The physical universe beyond the earth's atmosphere _____

10. The police conducted a raid on the criminals.
CLUE: Too dry or barren to support vegetation _____

11. Mary won a prize at the fair for the nicest floral arrangement.
CLUE: Arthropod with 6 jointed legs and body with head, thorax, & abdomen _____

12. The child drew a picture of a heart to give to his mother on Valentine's Day.
CLUE: The third planet from the sun _____

Hidden Anagrams: Animal Theme

Find the anagram of an animal hiding in each sentence.

1. Jessica and Aileen dipped their feet in the water as they walked along the shore.
CLUE: An equine _____

2. The newly paved road was closed because the tar was not dry.
CLUE: A rodent _____

3. The woman asked the butcher for a pork loin roast.
CLUE: A feline _____

4. The child tried to sneak a cookie from the cookie jar without his mother noticing.
CLUE: A limbless reptile _____

5. The girl asked to sit up front so she could hear the teacher.
CLUE: A gnawing animal with long hind legs in same family as a rabbit _____

6. The shoe store was having a big sale on winter boots.
CLUE: A pinniped _____

7. Jake went to the movies with his aunt and uncle.
CLUE: A saltwater fish in mackerel family _____

8. The artist prides herself on her exquisite use of color.
CLUE: An arachnid _____

9. The queen was seated in a throne made of gold.
CLUE: An insect in wasp family _____

10. Poseidon was the god of the sea in the mythology of the ancient Greeks.
CLUE: A canine _____

11. The ceiling was so low that the tall man had to bend down as he walked.
CLUE: A nocturnal bird of prey _____

12. The meteorologist said that most of the hail was about the size of a pea.
CLUE: A primate _____

Solutions*

*Optional Lists of Answers

Alphabetical lists of the answers are provided. These may be used to help solve the puzzle from the beginning, to assist those having difficulty, or not at all.

Animal Characteristics

A completed crossword puzzle with the following words:

Across:
- BIRD
- REPTILES
- CANINES
- ENDOSKELETON
- WARMBLOODED
- SPECIES
- VERTEBRATES
- AMPHIBIANS
- ARTHROPODS
- INVERTEBRATE
- HERBIVORES
- FELINES
- PINNIPED
- MOLLUSKS

Down:
- MARSUPIAL
- QUADRUPED
- CARNIVOR
- INSEECH (INSECTS / BRANCHIA)
- GILL
- METAMMO
- PREDATOR
- PRIMM
- FISH
- CTACACFAS
- EGGS
- EQUINES
- BOVINES
- OMNIVORE
- RODENT
- MAMMAL

Matter and Energy

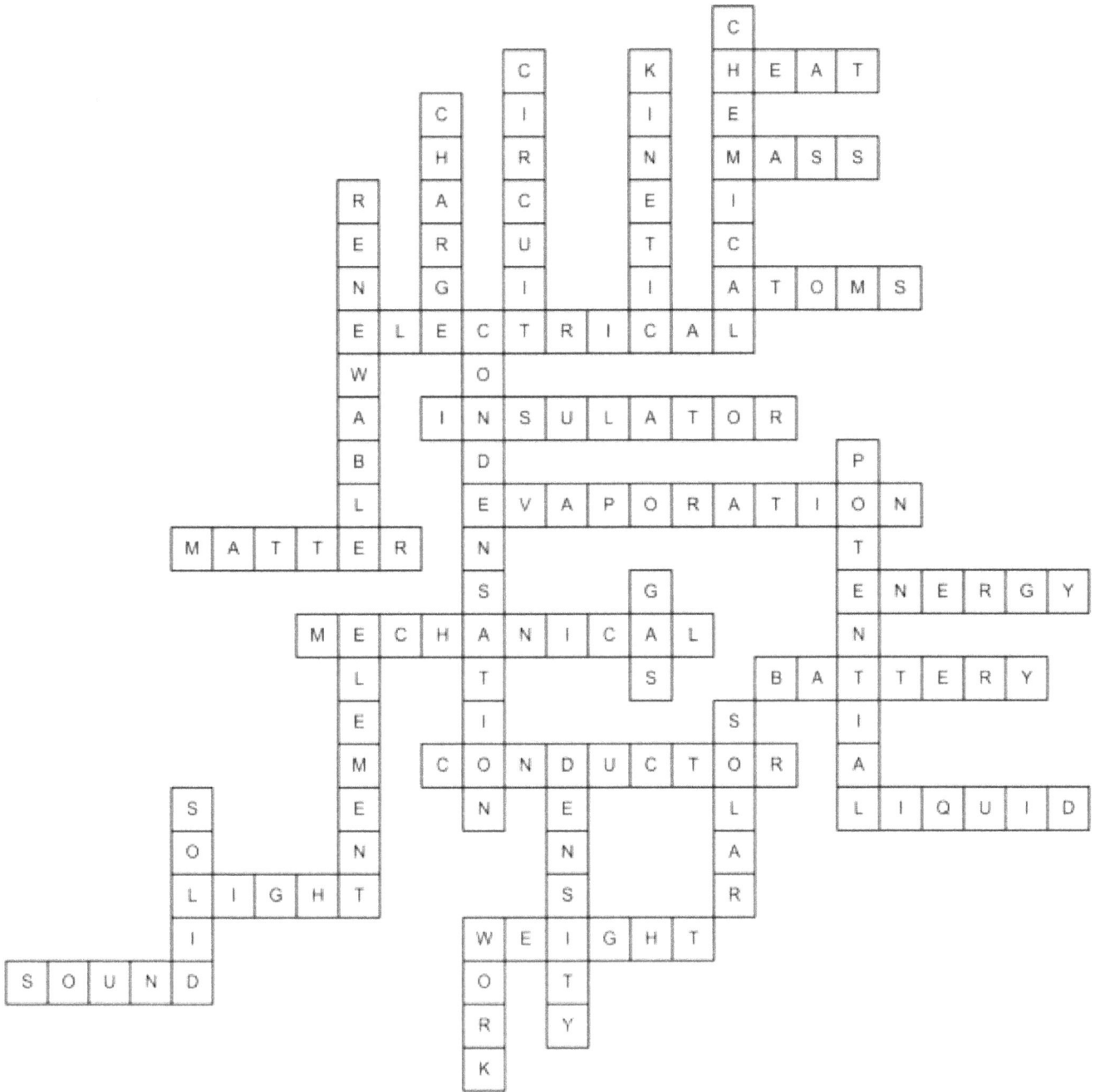

A completed crossword puzzle with the following answers:

HEAT, KINETIC, CHEMICAL, MASS, ATOMS, CIRCUIT, CHARGE, RENEWABLE, ELECTRICAL, CONDENSATION, INSULATOR, EVAPORATION, POTENTIAL, MATTER, ENERGY, MECHANICAL, GAS, BATTERY, ELEMENT, CONDUCTOR, POTENTIAL, SOLID, SOLAR, LIGHT, LIQUID, SOUND, WEIGHT, WORK, DENSITY

Forces and Motion

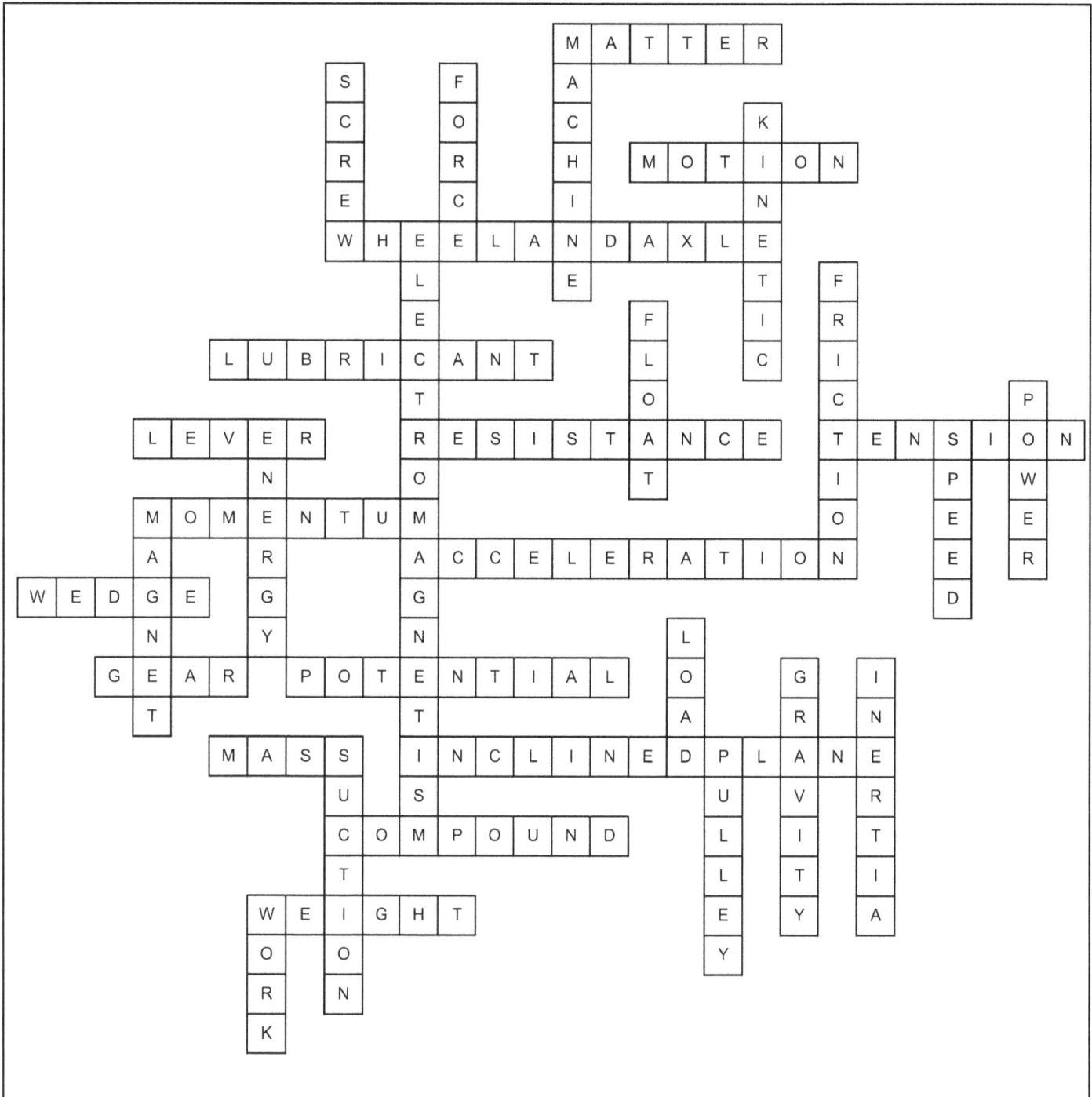

A crossword puzzle with the following answers filled in:

- MATTER
- SCREW
- FORCE
- MACHINE
- MOTION
- KINETIC
- WHEEL AND AXLE
- FRICTION
- LUBRICANT
- FLOAT
- LEVER
- RESISTANCE
- TENSION
- POWER
- MOMENTUM
- SPEED
- ACCELERATION
- WEDGE
- MAGNET
- ENERGY
- GEAR
- POTENTIAL
- LOAD
- GRAVITY
- INERTIA
- MASS
- INCLINED PLANE
- PUSH
- COMPOUND
- PULLEY
- WEIGHT
- WORK
- DOWN

The Human Body

A crossword puzzle grid containing the following words:

ARTERY, URINARY, SKELETAL, BLOOD, RESPIRATORY, VEINS, MUSCLES, BONES, CELL, SPINAL CORD, PULMONARY, KIDNEYS, TISSUE, SKULL, BRAIN, SYSTEM, HEART

Down words include: NERVOUS, DIGESTIVE, CIRCULATORY, JOINT, LARYNX, NOSE, LUNG, IMMUNE, STOMACH, EYE, GENE, TEETH, MARROW, EAR

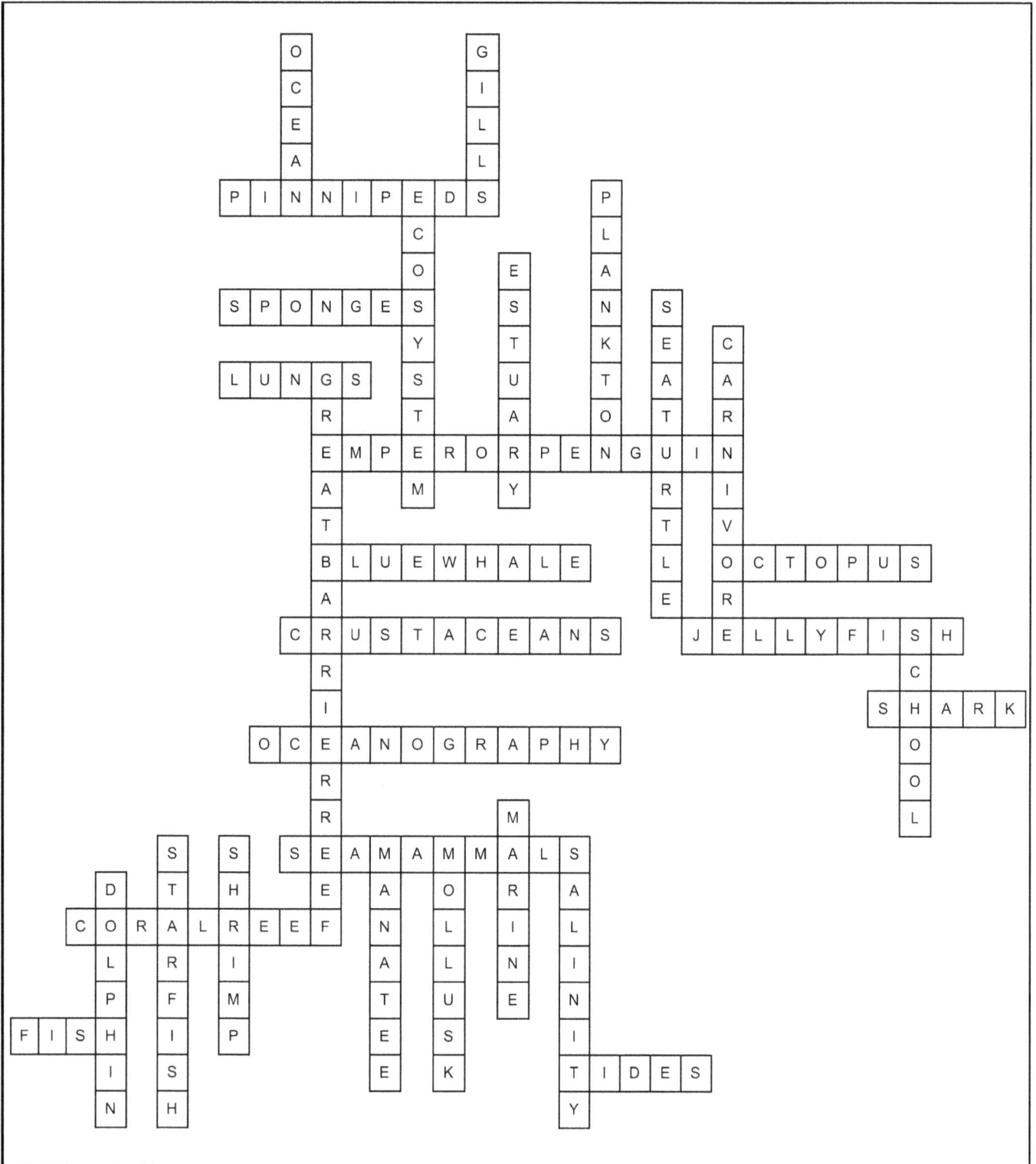

Marine Life

A completed crossword puzzle containing the following answers:

Across: PINNIPEDS, SPONGES, LUNGS, EMPEROR PENGUIN, BLUE WHALE, OCTOPUS, CRUSTACEANS, JELLYFISH, SHARK, OCEANOGRAPHY, SEA MAMMALS, CORAL REEF, FISH, TIDES

Down: OCEAN, GILLS, ECOSYSTEM, ESTUARY, PLANKTON, SEA TURTLE, CARNIVOR, CORAL SCHOOL, SEAMANATEE, MOLLUSK, MARINE BIOLOGY, DOLPHIN, STARFISH, SHRIMP

Our Solar System

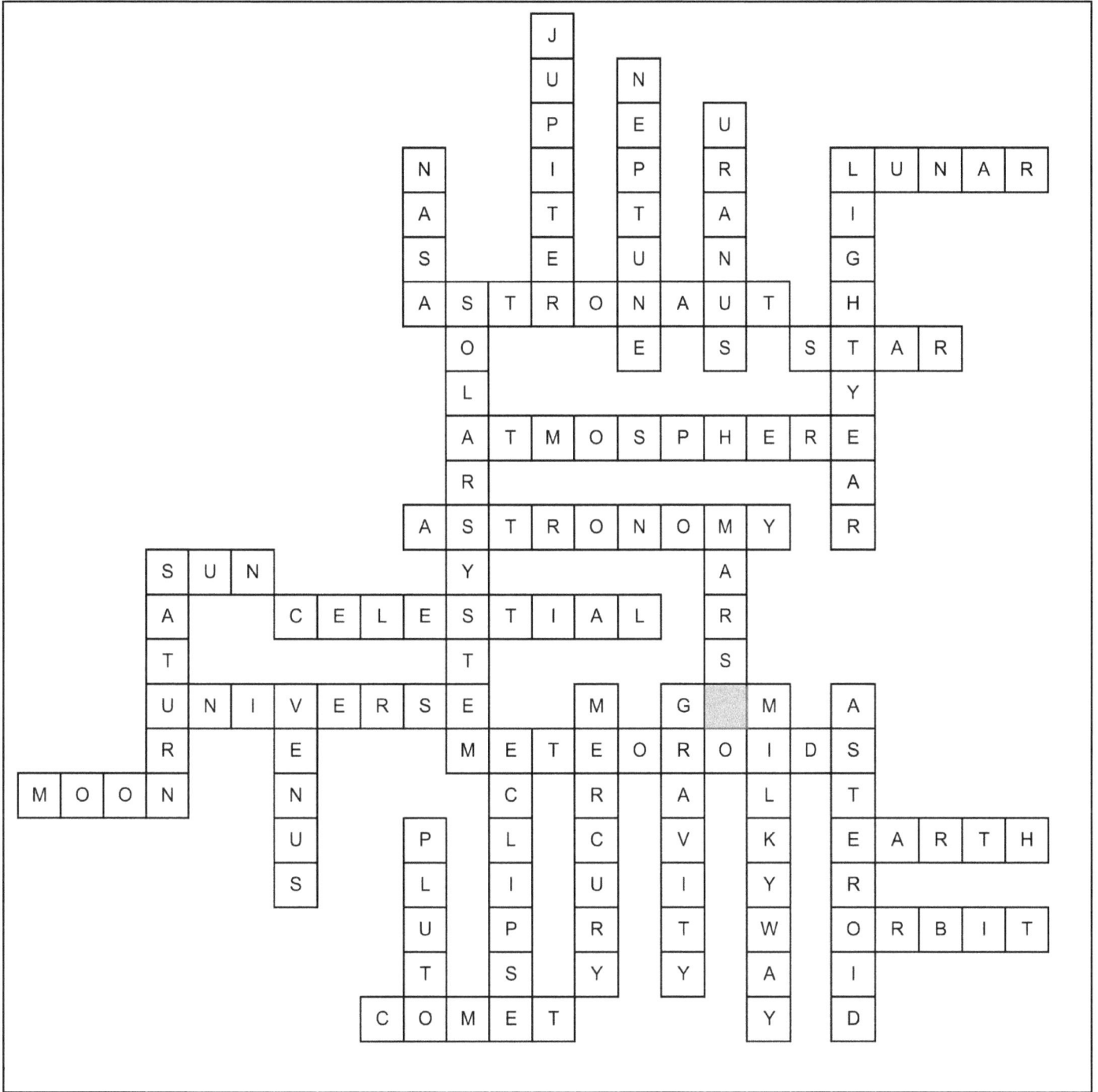

A completed crossword puzzle containing the following answers:

JUPITER, NEPTUNE, URANUS, LUNAR, NASA, LIGHT, ASTRONOMER, STAR, SOLAR, ATMOSPHERE, YEAR, ASTRONOMY, MARS, SUN, SATURN, CELESTIAL, SYSTEM, UNIVERSE, METEOR, MAGMA, METEOROIDS, MOON, VENUS, MERCURY, GRAVITY, MILKYWAY, EARTH, PLUTO, ELLIPSE, ORBIT, ASTEROID, COMET

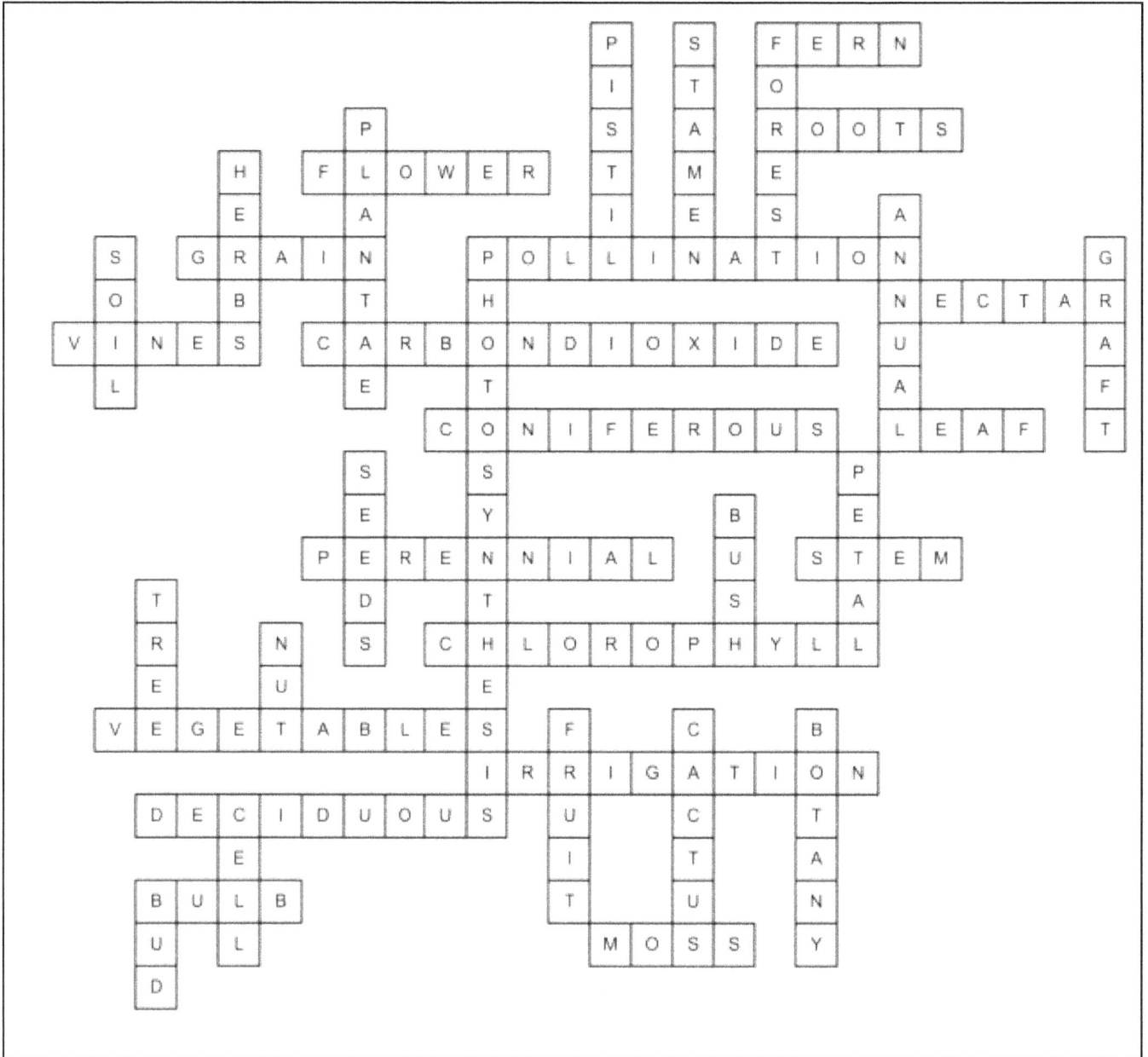

Plants

A completed crossword puzzle with the following words:

Across:
- FERN
- ROOTS
- FLOWER
- GRAIN
- POLLINATION
- NECTAR
- VINES
- CARBON DIOXIDE
- CONIFEROUS
- LEAF
- PERENNIAL
- STEM
- CHLOROPHYLL
- VEGETABLES
- IRRIGATION
- DECIDUOUS
- BULB
- MOSS

Down:
- PISTIL
- STAMEN
- FOREST
- PLANT
- HERBS
- SOIL
- ANNUAL
- GRAFT
- PHOTOSYNTHESIS
- SEEDS
- SYSTEM
- BUSH
- PEA
- TREE
- NUT
- FRUIT
- CACTUS
- BOTANY
- BUD

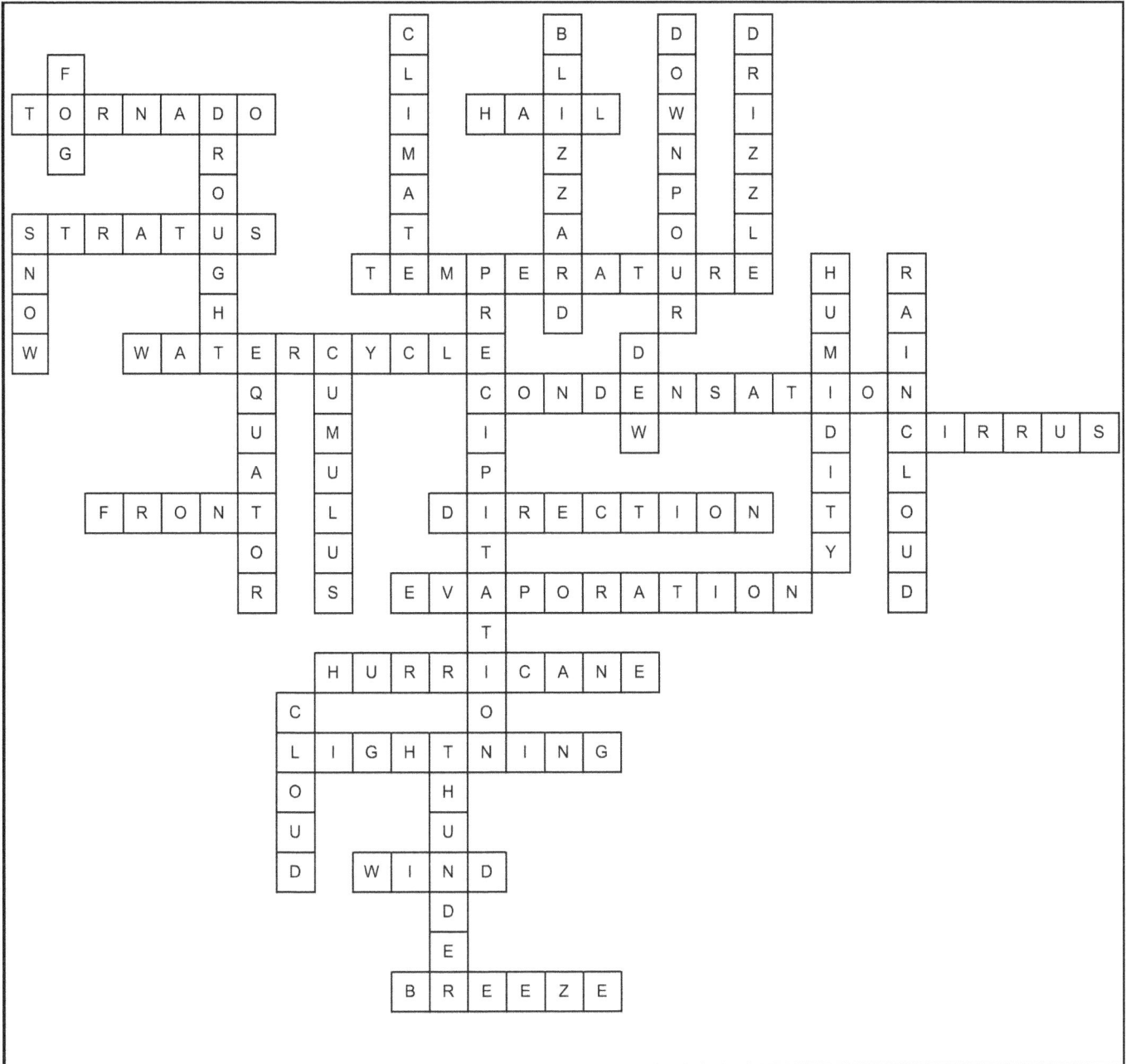

Weather

A completed crossword puzzle containing the following weather-related words:

Across and Down entries:

- TORNADO
- FOG
- STRATUS
- SNOW
- WATER CYCLE
- FRONT
- HAIL
- CLIMATE
- BLIZZARD
- TEMPERATURE
- CONDENSATION
- DIRECTION
- EVAPORATION
- HURRICANE
- LIGHTNING
- WIND
- BREEZE
- DOWNPOOL
- DRIZZLE
- HUMIDITY
- RAIN
- CIRRUS
- CLOUD
- CUMULUS
- PRECIPITATION
- FORECAST
- DEW

Earth Science

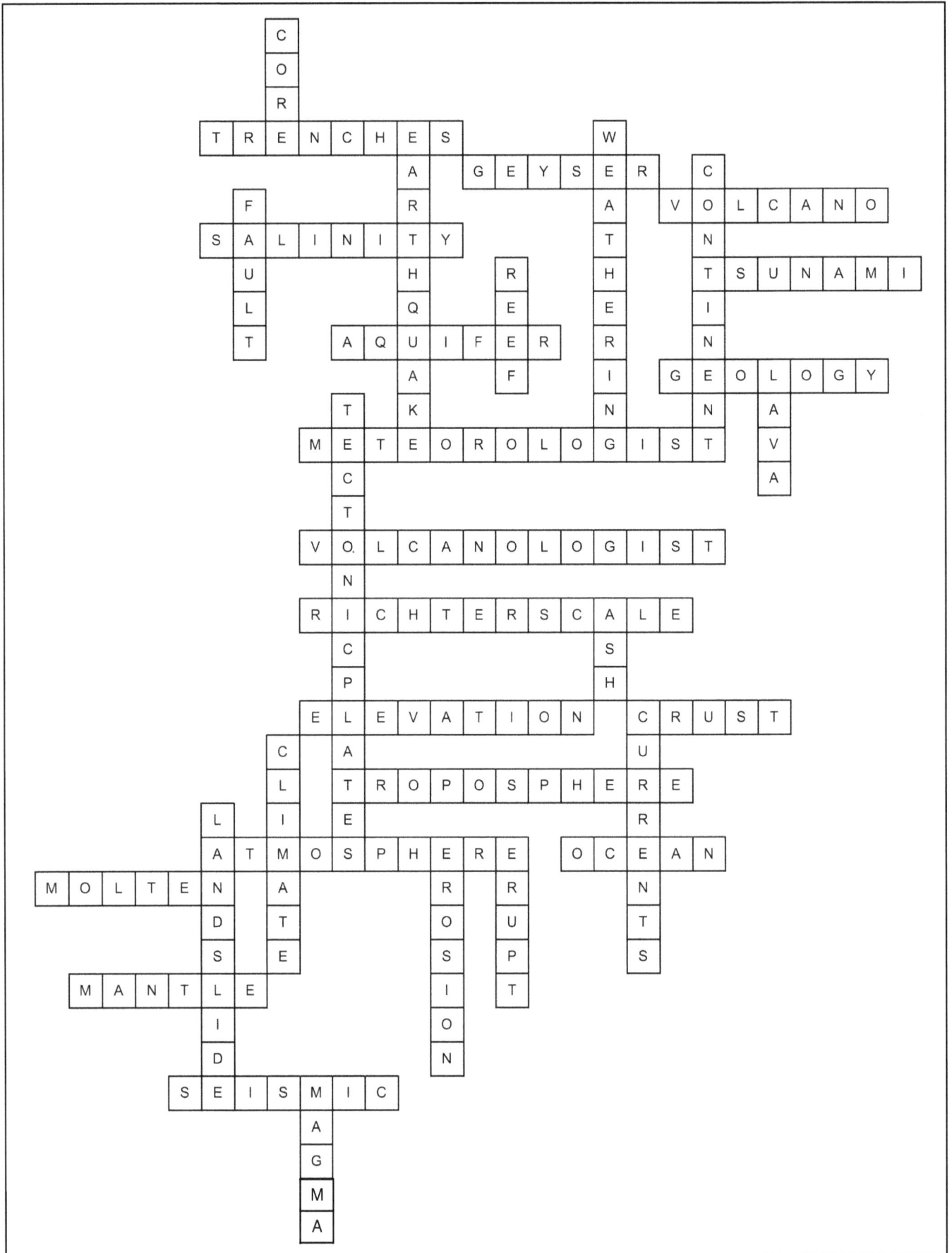

A completed crossword puzzle containing the following words:

CORE, TRENCHES, EARTHQUAKE, GEYSER, WEATHERING, VOLCANO, FAULT, SALINITY, TSUNAMI, REEF, AQUIFER, GEOLOGY, CONTINENTS, LAVA, METEOROLOGIST, TECTONIC, VOLCANOLOGIST, CONICPLATE(?), RICHTERSCALE, ASH, ELEVATION, CRUST, CLIMATE, PLATE, TROPOSPHERE, CURRENTS, LANDSLIDE, ATMOSPHERE, OCEAN, MOLTEN, CLIMATE, EROSION, ERUPT, MANTLE, SEISMIC, MAGMA

Rocks and Minerals

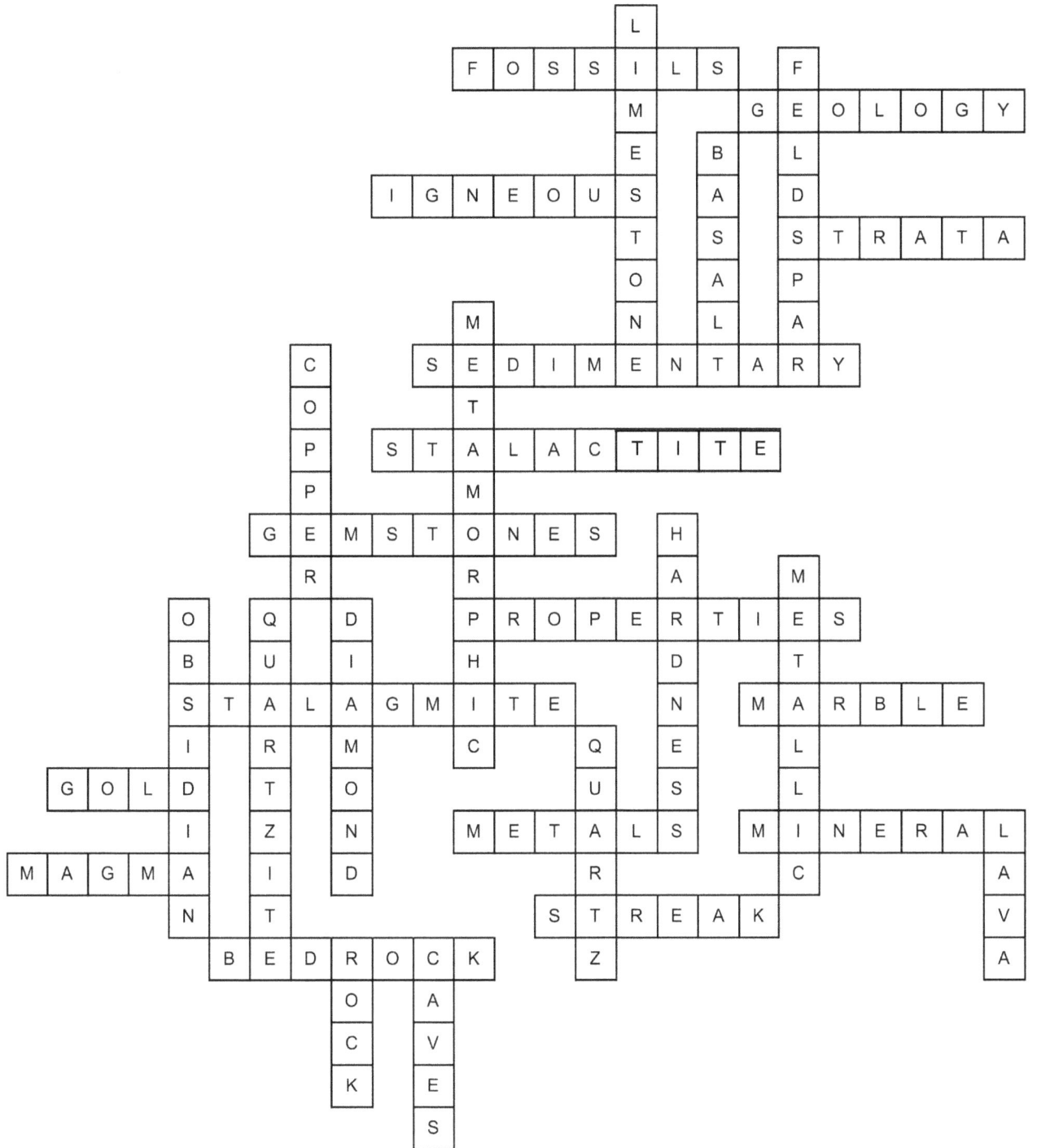

The completed crossword puzzle contains the following answers:

- FOSSILS
- GEOLOGY
- IGNEOUS
- STRATA
- SEDIMENTARY
- STALACTITE
- GEMSTONES
- PROPERTIES
- MARBLE
- STALAGMITE
- GOLD
- METALS
- MINERAL
- MAGMA
- STREAK
- BEDROCK

Down answers include: LIMESTONE, FELDSPAR, BASALT, COPPER, METAMORPHIC, OBSIDIAN, QUARTZITE, DIAMOND, PHOSPHATE, HARDNESS, QUARTZ, METALLIC, ROCKS, CAVES, LAVA

Classification of Living Things Word Search

	1	2	3	4	5	6	7	8	9	10	11	12	13	14	15	16	17	18
1	W	U	C	M	T	T	U	S	Z	O	A	N	A	H	I	L	S	C
2	F	C	L	L	A	I	P	L	R	D	Z	Y	H	O	R	M	D	M
3	S	A	Y	C	A	E	G	V	E	R	T	E	B	R	A	T	E	H
4	K	I	S	G	C	S	Z	E	N	I	L	E	F	O	N	N	F	S
5	A	Y	R	I	I	D	S	V	R	C	T	A	J	O	Z	A	O	X
6	E	N	E	G	N	E	X	L	A	R	E	H	T	N	A	P	K	B
7	S	S	I	Z	I	B	R	I	A	G	P	T	K	J	N	H	F	O
8	C	S	N	M	A	T	A	O	E	M	O	T	A	C	G	I	B	Q
9	I	Y	K	L	A	E	A	V	R	M	W	W	K	O	W	M	K	
10	Z	N	N	U	A	L	D	R	D	I	Z	A	A	J	T	W	C	I
11	R	Q	K	H	F	I	I	E	E	N	N	R	M	A	D	A	R	N
12	M	E	W	H	L	O	R	A	D	H	O	R	T	B	M	N	Z	G
13	E	G	W	E	S	O	B	Y	A	V	T	A	A	U	E	K	P	D
14	E	K	F	U	J	X	L	N	I	K	D	N	L	C	K	A	B	O
15	A	A	N	K	P	I	I	N	X	R	D	Y	A	U	L	I	H	M
16	P	E	K	X	M	M	R	E	O	F	H	U	N	P	Z	U	N	F
17	G	T	S	A	A	A	U	H	V	P	A	I	L	A	M	M	A	M
18	M	Z	F	L	C	I	C	M	T	V	A	U	U	A	F	Y	C	H

The terms below are listed with their starting row and column.

ANIMAL 13:9

ANIMALIA 5:1

BIG CAT 8:17

CARNIVORA 18:5

CARNIVORE 14:14

CAT 3:4

CHORDATA 18:7

CLASS 1:3

FAMILY 18:3

FELIDAE 14:3

FELINE 4:13

GENUS 17:1

KINGDOM 9:18

MAMMAL 11:13

MAMMALIA 17:18

ORDER 8:11

PANTHERA 6:16

PANTHERA TIGRIS 16:14

PHYLUM 17:10

SPECIES 1:8

TIGER 1:5

VERTEBRATE 3:8

Hidden Anagrams: Science Theme

Find the anagram of a science term hiding in each sentence.

1. A deep moat surrounded the castle.
CLUE: Smallest unit of an element that retains the chemical properties **moat - atom**

2. The teen searched in the couch cushions for the remote control.
CLUE: Small body of matter from outer space that enters the earth's atmosphere **remote - meteor**

3. The shelf seemed to sag in the middle because of the heavy weight placed on it.
CLUE: Substance that is like air and has no fixed shape **sag - gas**

4. The boy began to lament his decision not to attend the party.
CLUE: Beyond the Earth's crust **lament - mantle**

5. The speech given during the assembly was very topical.
CLUE: Relating to or using light **topical - optical**

6. The guest asked her hostess for the recipe for the clam dip.
CLUE: Pleasantly free from wind **clam - calm**

7. The young child was learning how to tie the laces on his shoes.
CLUE: Small, thin horny or bony plate protecting skin of fish and reptiles **laces - scale**

8. The chef used many different oils, but olive oil was his favorite.
CLUE: The upper layer of earth in which plants grow **oils - soil**

9. The directions on the treasure map said to take three paces to the right.
CLUE: The physical universe beyond the earth's atmosphere **paces - space**

10. The police conducted a raid on the criminals.
CLUE: Too dry or barren to support vegetation **raid - arid**

11. Mary won a prize at the fair for the nicest floral arrangement.
CLUE: Arthropod with 6 jointed legs and body with head, thorax, & abdomen **nicest- insect**

12. The child drew a picture of a heart to give to his mother on Valentine's Day.
CLUE: The third planet from the sun **heart - Earth**

Hidden Anagrams: Animal Theme

Find the anagram of an animal hiding in each sentence.

1. Jessica and Aileen dipped their feet in the water as they walked along the shore.
CLUE: An equine __**shore - horse**__

2. The newly paved road was closed because the tar was not dry.
CLUE: A rodent __**tar - rat**__

3. The woman asked the butcher for a pork loin roast.
CLUE: A feline __**loin - lion**__

4. The child tried to sneak a cookie from the cookie jar without his mother noticing.
CLUE: A limbless reptile __**sneak - snake**__

5. The girl asked to sit up front so she could hear the teacher.
CLUE: A gnawing animal with long hind legs in same family as a rabbit __**hear - hare**__

6. The shoe store was having a big sale on winter boots.
CLUE: A pinniped __**sale - seal**__

7. Jake went to the movies with his aunt and uncle.
CLUE: A saltwater fish in mackerel family __**aunt - tuna**__

8. The artist prides herself on her exquisite use of color.
CLUE: An arachnid __**prides - spider**__

9. The queen was seated in a throne made of gold.
CLUE: An insect in wasp family __**throne - hornet**__

10. Poseidon was the god of the sea in the mythology of the ancient Greeks.
CLUE: A canine __**god -dog**__

11. The ceiling was so low that the tall man had to bend down as he walked.
CLUE: A nocturnal bird of prey __**low - owl**__

12. The meteorologist said that most of the hail was about the size of a pea.
CLUE: A primate __**pea - ape**__

Optional Lists of Words and Terms

These lists are provided for your convenience. If a puzzle is used as an introduction or just for fun, you might want to provide the list of words. On the other hand, if the puzzle is being done in lieu of a quiz, you might choose not to utilize them. In either case, solutions to the puzzles are provided.

Animal Characteristics

amphibians   arachnids   arthropods   bird   bovines   canines   carnivore   cetaceans   eggs   endoskeleton   equines   felines   fish   gills   herbivores   insect   invertebrate   mammal   marsupials   metamorphosis   mollusks   omnivores   pinniped   predator   primates   quadruped   reptiles   rodents   species   vertebrates   warm blooded

---

Matter and Energy

atoms   battery   charge   chemical   circuit   condensation   conductor   density   electrical   element   energy   evaporation   gas   heat   insulator   kinetic energy   light   liquid   mass   matter   mechanical   potential energy   renewable   solar   solid   sound   weight   work

---

Forces and Motion

acceleration   compound   electromagnetism   energy   float   force   friction   gear   gravity   inclined plane   inertia   kinetic   lever   load   lubricant   machine   magnet   mass   matter   momentum   motion   potential   power   pulley   resistance   screw   speed   suction   tension   wedge   weight   wheel and axle   work

---

The Human Body

artery   blood   bones   brain   cell   circulatory   digestive   ear   eye   gene   heart   immune   joints   kidneys   larynx   lungs   muscles   nervous   nose   pulmonary   respiratory   skeletal   skull   spinal cord   stomach   system   teeth   tissue   urinary   veins

---

Marine Life

blue whale   carnivore   coral reef   crustaceans   dolphin   ecosystem   emperor penguin   estuary   fish   gills   Great Barrier Reef   jellyfish   lungs   manatee   marine   mollusk   ocean   oceanography   octopus   pinnipeds   plankton   salinity   school   sea mammals   sea turtle   shark   shrimp   sponges   starfish   tides

---

Our Solar System

asteroid　　astronaut　　astronomy　　atmosphere　　celestial
comet　　Earth　　eclipse　　gravity　　Jupiter　　light year
lunar　　Mars　　Mercury　　meteoroids　　Milky Way
moon　　NASA　　Neptune　　orbit　　Pluto　　Saturn
solar system　　star　　sun　　universe　　Uranus　　Venus

Plants

annual　　botany　　bud　　bulb　　bush　　cactus　　carbon dioxide　　cell
chlorophyll　　coniferous　　deciduous　　fern　　flower　　forest　　fruit
graft　　grain　　herbs　　irrigation　　leaf　　moss　　nectar　　nut
perennial　　petal　　photosynthesis　　pistil　　Plantae　　pollination
roots　　seeds　　soil　　stamen　　stem　　tree　　vegetables　　vines

Weather

blizzard　　breeze　　cirrus　　climate　　cloud　　condensation
cumulus　　dew　　direction　　downpour　　drizzle　　drought
equator　　evaporation　　fog　　front　　hail　　humidity　　hurricane
lightning　　precipitation　　rain cloud　　snow　　stratus
temperature　　thunder　　tornado　　water cycle　　wind

Earth Science

aquifer　　ash　　atmosphere　　climate　　continent　　core　　crust
currents　　earthquake　　elevation　　erosion　　erupt　　fault　　geology　　geyser
landslide　　lava　　magma　　mantle　　meteorologist　　molten
ocean　　reef　　Richter scale　　salinity　　seismic　　tectonic plates　　trenches
troposphere　　tsunami　　volcano　　volcanologist　　weathering

Rocks and Minerals

basalt　　bedrock　　caves　　copper　　diamond　　feldspar
fossils　　gemstones　　geology　　gold　　hardness　　igneous　　lava
limestone　　magma　　marble　　metallic　　metals　　metamorphic
mineral　　obsidian　　properties　　quartz　　quartzite　　rock
sedimentary　　stalactite　　stalagmite　　strata　　streak